نمبریں دی کہانی

THE NUMBER STORY

SMALL BOOK ONE

ENGLISH - SARAIKI

Numbers Teach Children
Their Number Names

written and illustrated by

MISS ANNA

Early Reader Edition of *The Number Story 1*
Bronze Medal Winner, 2016 Wishing Shelf Book Award

Library of Congress Control Number: 2018902040

Names: Miss Anna, author.
Title: Number story : numbers teach children their number names / Miss Anna.
Description: Portland, OR: Lumpy Publishing, 2018.
Identifiers: ISBN 978-1-945977-97-8| LCCN 2018902040
Summary: The pictures and rhymes present stories which introduce numbers 0-10.
Subjects: LCSH Numeration—English--Saraiki--Pictorial works--Juvenile literature. | BISAC JUVENILE NONFICTION /
Languages: English--Saraiki
Classification: LCC QA141.3 .M57 2018 | DDC 513—dc23

Publisher: Lumpy Publishing
Website: www.missannabooks.com
Email: missanna@missannabooks.com

Paperback: ISBN 978-1-945977-97-8
Printed in the U.S.A.   1 3 5 7 9 10 8 6 4 2

Want to learn our number names?

اساڈے نمبریں دے ناں سکھنڑ
چاہندے او؟

It is very easy and a lot of fun!

اے بھوں آسان تے ایندے وچ کافی تفریح اے!

Say-along our little jingle

اساڈی چھوٹی جئیں کہانی دے نال گانوو!

starting from Number One!

نمبر هڪ کنوں شروع ٿيندے!

ONE looks like my one finger.

ہک میڈی ہک انگل آلی کار لگدے ۔

ONE!
هک!

2

TWO   trails a tail.

دو پوچھڑ گھلیندے ۔

هک پوچھڑ! A TAIL!

3

THREE   has bumps.

ترآے دے ابھار ہن۔

ابهاريں آلا!

4

FOUR   carries a sail.

چار بادبان چئی ودے ۔

A SAIL!
یک بادبان!

5

FIVE   is a racing track.

پنج ہک دوڑ دا رستہ اے ۔

VROOM
أوقفي

6

SIX   curves like a snail.

چھی گھونگھے آلی کار کُبا۔

ہک گھونگھا!

7

SEVEN has a sharp angle.

ست دا تِکھا کونہ ہے ۔

BE CAREFUL! IT'S SHARP!
دھیان کریں! اے وڈا تیز اے!

EIGHT   is rollercoaster rails.

اٹھ ہک رولر کوسٹر دی پٹڑی۔

یپی!

YIPPEE!

9

NINE   is a bubble on a stick.

نائوں چھڑی تے ببل اے ۔

A BUBBLE!

هك يبل!

10
TEN   is an eye of a whale.
ڈھا ویہل دی ہک اکھ۔

WINK!
اكه مار!
HELLO! هيلو!

And
تے
0
ZERO   is an empty pail.
صفر ہک خالی ڈول اے ۔

IT'S
EMPTY!
اے خالی اے!

Thank you for playing with us today.

We had a lot of fun too!

اج اساڈے نال کھیڈنڑ کیتے توآڈا بھوں شکریہ۔

اج اساکوں وڈی چس آئی اے۔

We are your Number friends,
Zero to Ten,
Who will be here for you~

اساں توآڈے نمبر یار آؤں
صفر کنوں ڈھا تئیں۔
اساں توآڈے نال اتھاں ہمیشہ رہسوں۔

Bye-bye now!
See you again soon!

ہنڑ وت بائے بائے!
ولا ملسوں!

The Numbers are *SINGING* too!

To sing-a-long, look for Miss Anna Number Story
at your favorite music store like iTUNES.

MP3

Numbers 0-10
IDENTIFYING & COUNTING

Numbers 11-20
& Ordinals
first, second, third...

Numbers 0-100
& Place Values
ones, tens, hundreds...

About Clocks
& Telling Time
hours, minutes, seconds

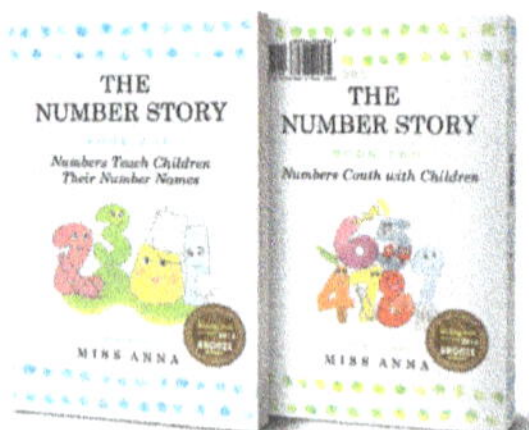

Number Story 1 & 2
isbn: 978-0-996216-48-7

Number Story 3 & 4
isbn: 978-1-945977-01-5

Number Story 5 & 6
isbn: 978-1-945977-06-0

Number Story 7 & 8
isbn: 978-1-949320-40-4

For more Miss Anna books to love,
visit us at

www.missannabooks.com

Numbers are working hard all over the world!
Come Travel the World with Us!